Making Android Accessories with IOIO

Simon Monk

O'REILLY®

Beijing · Cambridge · Farnham · Köln · Sebastopol · Tokyo

Making Android Accessories with IOIO
by Simon Monk

Published by O'Reilly Media, Inc., 1005 Gravenstein Highway North, Sebastopol, CA 95472.

O'Reilly books may be purchased for educational, business, or sales promotional use. Online editions are also available for most titles (*http://my.safaribooksonline.com*). For more information, contact our corporate/institutional sales department: (800) 998-9938 or *corporate@oreilly.com*.

Editors: Andy Oram and Mike Hendrickson
Production Editor: Kristen Borg
Proofreader: O'Reilly Production Services
Cover Designer: Karen Montgomery
Interior Designer: Ron Bilodeau and Edie Freedman
Illustrator: Robert Romano

February 2012:   First Edition.

Revision History for the First Edition:
   February 13, 2012   First release
See *http://oreilly.com/catalog/errata.csp?isbn=9781449323288* for release details.

ISBN: 978-1-449-32328-8
[LSI]
1328898864

Contents

Preface

Android phones are a great platform for developing apps, but sometimes it is nice if those apps go beyond the built-in hardware of the phone and connect to some homemade electronics.

The IOIO board allows you to do just that, and this book will show you how to use the IOIO board and interface it to various different electronic modules and components.

These techniques involved in using IOIO are illustrated in example projects. These projects are:

- An intruder alarm that uses your phone to send an SMS text message when movement is detected by its PIR sensor.
- A Bluetooth temperature logger that records temperatures onto the SD card of your phone.
- An 8x8 LED Matrix display that will display animations and is controlled by your phone.
- A Bluetooth rover that you can control from your Android phone.

What You Will Need

For all the projects, you will need an Android phone running Android 2.1 or later, and of course, an IOIO board.

Each project also requires some additional parts, and these are listed along with order codes for US and international component suppliers.

The projects are of various levels of difficulty and all require a little soldering, so you will also need a soldering iron.

How to Use this Book

You need to read Chapter 1 to get started, but then you can pick and choose from the remaining project chapters. All the code for the projects is available at *http://www.ioiobook.com*.

Conventions Used in This Book

The following typographical conventions are used in this book:

Italic

> Indicates new terms, URLs, email addresses, filenames, and file extensions.

`Constant width`

> Used for program listings, as well as within paragraphs to refer to program elements such as variable or function names, databases, data types, environment variables, statements, and keywords.

`Constant width bold`

> Shows commands or other text that should be typed literally by the user.

`Constant width italic`

> Shows text that should be replaced with user-supplied values or by values determined by context.

 CAUTION: This icon indicates a warning or caution.

Using Code Examples

This book is here to help you get your job done. In general, you may use the code in this book in your programs and documentation. You do not need to contact us for permission unless you're reproducing a significant portion of the code. For example, writing a program that uses several chunks of code from this book does not require permission. Selling or distributing a CD-ROM of examples from O'Reilly books does require permission. Answering a question by citing this book and quoting example code does not require permission. Incorporating a significant amount of example code from this book into your product's documentation does require permission.

We appreciate, but do not require, attribution. An attribution usually includes the title, author, publisher, and ISBN. For example: "*Making Android Accessories with IOIO* by Simon Monk (O'Reilly). Copyright 2012 Simon Monk, 978-1-449-32328-8."

If you feel your use of code examples falls outside fair use or the permission given above, feel free to contact us at *permissions@oreilly.com.*

Safari® Books Online

 Safari Books Online is an on-demand digital library that lets you easily search over 7,500 technology and creative reference books and videos to find the answers you need quickly.

With a subscription, you can read any page and watch any video from our library online. Read books on your cell phone and mobile devices. Access new titles before they are available for print, and get exclusive access to manuscripts in development and post feedback for the authors. Copy and paste code samples, organize your favorites, download chapters, bookmark key sections, create notes, print out pages, and benefit from tons of other time-saving features.

O'Reilly Media has uploaded this book to the Safari Books Online service. To have full digital access to this book and others on similar topics from O'Reilly and other publishers, sign up for free at *http://my.safaribooksonline.com*.

How to Contact Us

Please address comments and questions concerning this book to the publisher:

> O'Reilly Media, Inc.
> 1005 Gravenstein Highway North
> Sebastopol, CA 95472
> 800-998-9938 (in the United States or Canada)
> 707-829-0515 (international or local)
> 707-829-0104 (fax)

We have a web page for this book, where we list errata, examples, and any additional information. You can access this page at:

> *http://shop.oreilly.com/product/0636920024668.do*

To comment or ask technical questions about this book, send email to:

> *bookquestions@oreilly.com*

For more information about our books, courses, conferences, and news, see our website at *http://www.oreilly.com*.

Find us on Facebook: *http://facebook.com/oreilly*

Follow us on Twitter: *http://twitter.com/oreillymedia*

Watch us on YouTube: *http://www.youtube.com/oreillymedia*

Acknowledgments

I thank Linda for giving me the time, space, and support to write this book, and for putting up with the various messes my projects create around the house.

Thanks to Ytai Ben-Tsvi, the originator of IOIO, for doing such a good job on the platform and his most useful comments on the book during its writing.

Nathan and Aaron at Sparkfun kept me supplied in IOIOs, and I thank them for their help and encouragement.

Finally, I would like to thank Andy Oram, Mike Hendrickson, and everyone at O'Reilly who has had a hand in producing this book.

1/Getting Started with IOIO

IOIO (pronounced *YoYo*) is an input/output board for Android phones and tablets. It allows you to attach electronic devices to your Android phone using the USB connection. If you have a IOIO with the latest firmware, you can also communicate over Bluetooth if you attach a Bluetooth adapter to the IOIO.

In this chapter we will look at how to set up your computer and Android phone to use the IOIO and use the sample application that comes with the IOIO software to turn the "status" LED on and off for your phone.

What is IOIO?

IOIO (Figure 1-1) is a product produced and sold by SparkFun (among others). It contains a PIC microcontroller and USB interface and a few other glue components to regulate the supply voltage etc. The whole project, both hardware and software is open source.

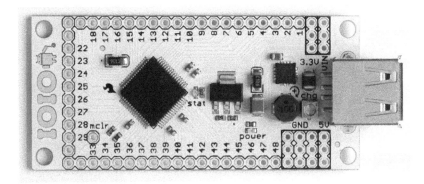

Figure 1-1. *An IOIO board*

To make use of an IOIO, you will need:

- An Android phone with USB lead
- A Windows, Mac, or Linux computer
- A power supply 5-15V DC offering at least 500mA

Since the IOIO comes without any sockets soldered to its connectors, we will attach some to it and make a power lead for it. To do this, you will also need the parts listed in Table 1-1.

Table 1-1. *Parts bin*

Quantity	Description	SparkFun SKU	Farnell code	Newark code
1	IOIO board	DEV-10748		
2	SIL Header socket strip	PRT-00115	1126603	52K3454
1	2.1mm Power socket	PRT-00119	1217038	97K6459
2	single header pins	PRT-00116	1097954	93K5128
	short lengths of red and black multi-core wire			
1	9V power supply	TOL-00298	1354815	97F7919

Android Phone

IOIO will work with a wide range of Android phones. Most Android phones with Android 1.5 or later will work with this board. Any new phone that you buy will have at least Android 2.1 on it.

The type of USB connection that you need is USB client. This is what the vast majority of regular Android phones will have. This takes the form of a little USB-micro B connector. However, some newer Android devices have USB host connection that usually looks like a regular USB socket such as you would find on a desktop computer. This type of connection is not compatible with IOIO.

When you use IOIO, you write the program on your computer and then transfer it to your phone using the USB connection. You can then unplug the USB lead from your computer and plug it into the IOIO so that your phone is now connected to the IOIO board. There is no actual programming of the IOIO board itself. The program runs on the phone, which communicates with the IOIO over USB or Bluetooth.

Computer

IOIO uses the same Integrated Development Environment that Google recommends for Android development—Eclipse. Eclipse isn't required for either Android or IOIO, but it is the most common software used to create apps for them. Eclipse and the other software that you need are all available for Windows, Mac, and Linux.

Eclipse is fairly resource-hungry, so you will need a reasonably modern computer, or it will be slow and annoying to use. I use a 2.5GHz dual core Mac with 4GB and it works absolutely fine.

Power Supply

The IOIO does not take power from USB. So if you connect it to your phone with the USB lead, the power light on the IOIO will not illuminate.

The phone expects to receive charging power from the IOIO as if it were plugged into your computer, so you need to connect a power supply to the IOIO board. This may be a plug-in power supply or could be as simple as a small 9V battery. Although if you are connecting through USB rather than Bluetooth, a small 9V battery will not last long, as the charging current to the phone will soon empty it.

The projects in the following chapters use a mixture of power adapters and batteries.

None of the projects in this book, are very power-hungry and a 500mA (5W) supply is enough. However, the IOIO possesses a high-power voltage regulator, which means for more demanding projects, it can provide up to 1.5A at 5V. So, if you plan to use your IOIO to control high-power devices like motors and high-power LEDs, you may wish to buy say a 20W power supply. The rover project in Chapter 5 uses low-power motors, and is in any case battery-powered.

If you are looking for a power connector on the board, I'm sorry to say there isn't one. We will need to do some soldering to the board to make the power connections. Alternatively, there is an area on the back of the board, behind the USB connector where a surface mount JST connector can be soldered. SparkFun supplies such a connector (PRT-08612) as well as a power socket adaptor (TOL-08734), but they are not necessary for the projects in this book.

We are going to solder sockets to the board so that we can do most of our later project work with little or no soldering.

Preparing Your IOIO Board

In this section, we will prepare the IOIO board to be used by the various projects in this book. To do this, you will need to buy the items in Table 1-1.

When you get your IOIO board, it will be completely naked with no connectors or visible means of connecting electronics or even a power supply. In the various projects in this book, we will need to make both power and electronic input and output connections. To do this, we must find a way of attaching wires to the connector pads.

A convenient way to do this is to solder header strips to the two long sides of the IOIO. The projects in this book use only the first 20 connectors on each side and none of the connectors on the end. Figure 1-2 shows the board with the header sockets in place.

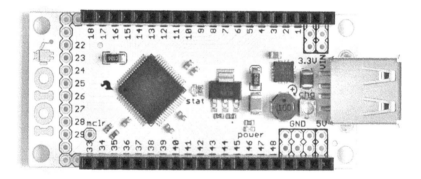

Figure 1-2. *An IOIO board with header sockets*

To solder the connectors, put them in place, then turn the board upside down so that its weight holds the sockets in position. Making sure that the sockets are straight, solder each connector in turn. When done, carefully check that there are no solder bridges between connectors.

For the power connection, we are going to make a short lead that has two header pins on one end and a standard 2.1mm low voltage power supply socket on the other end. This will allow us to drive the IOIO from a low-cost wall-wart type power supply (Figure 1-3).

Figure 1-3. *A power lead for the IOIO*

The IOIO board has the Vin pin and GND pins on opposite sides of the board, across which power should be provided. This means that our lead will need to separate at the board end. Twisting the wires together helps neaten the lead.

It is a good idea to use red wire for the positive connection to the center connection of the 2.1mm socket and a black lead from the outside connector of the socket to the GND pin. Notice that on the IOIO board, we have a choice of three GND sockets. Any one will do when connecting up the power.

Having got this far, we can at least check that our board will power up, by connecting the external power supply. You should find that the "Power" LED will light, and if you connect a phone to the USB cable, you should see that its charging LED will light, indicating that the IOIO is actually supplying power to the phone.

If the phone does not start charging, then you can use a small screwdriver to adjust the trim-pot just behind the USB socket. This controls how much current is supplied to the phone. If it is set too low, then the phone may not detect that the IOIO is attached.

WARNING: Most power supplies make the center pin of their 2.1mm plug the positive connection, but some adaptors, especially in the music world, are the other way around. So check the polarity before connecting up.

Installation

If you do not want to set your computer up to build your own IOIO apps, the IOIO apps used in the projects in this book are all available, ready built for download from the book's website [*http://www.ioiobook.com*].

But if you want to write your own apps for IOIO or want to get a better understanding of how the apps work, you will need to follow the instructions below. You effectively need to install everything that you need for Android software development, and then some extra code specific to making apps that use IOIO.

Overall, the steps involved are:

- Install Java.
- Install Eclipse.
- Install the Android SDK.
- Install the Eclipse Android ADT Plugin.
- Import the IOIO library and sample apps.

To install a suitable Java environment and Eclipse, see the instructions at [*http://wiki.eclipse.org/Eclipse/Installation*].

Once Eclipse is installed, install the Android SDK by following the instructions at [*http://developer.android.com/sdk*].

Instructions for installing the ADT Plugin for Eclipse can be found at [*http:// developer.android.com/sdk/eclipse-adt.html*].

Once the ADT Plugin has been installed, you will need to tell Eclipse about the location of the Android SDK. To do this, open the Preferences panel in Eclipse and select Android on the lefthand side. In the SDK field location, browse to the root directory of the ADT that you just installed (Figure 1-4).

When we installed the ADT earlier, we installed only the basic framework. We now need to install packages for the Android platform versions that we want to use. You should at least install the platform that matches your phone version. Normally, any applications that you build with an earlier version of Android will still work on a phone with a more recent version. Select the version of Android that matches that of your phone.

Platforms are added using a tool called the Android SDK and AVD Manager. This is launched from the Window menu in Eclipse, or by typing "android" from the Linux or Mac command lines. As you can see from Figure 1-5, the author has quite a few platforms installed. You can install as many as you like, because they will not conflict with each other.

If you want to use IOIO with the new Android Open Accessory framework, you will also need to install SDK Platform Android 2.2.2, API 10. Click on Available Packages to find packages to install. However, this is not necessary for the projects in this book.

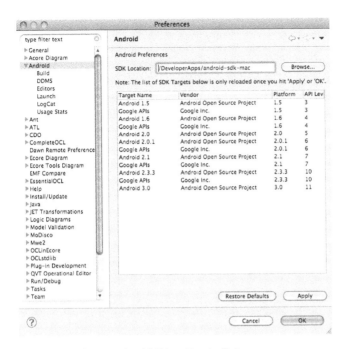

Figure 1-4. *Setting the ADT location in Eclipse*

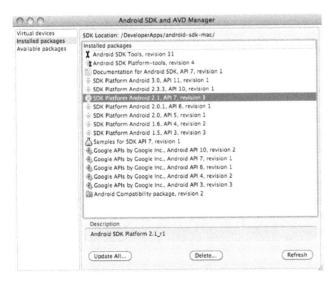

Figure 1-5. *Installing Android platforms in Eclipse*

Once your basic Android development setup is complete, you need to import all the IOIO sample projects and library from within Eclipse.

The first step is to download the zip file containing IOIOLib and the examples. You can find this at [*https://github.com/ytai/ioio/wiki/Downloads*].

Download the latest version. Version 3.10 and later have support for Bluetooth. The zip file will be called something like *App-IOIO0310.zip*.

Next, from Eclipse, right-click in the Project Explorer area and select Import, then General and Existing Projects into Workspace (Figure 1-6). Click the radio button for Select Archive File. Then browse to the zip file you downloaded.

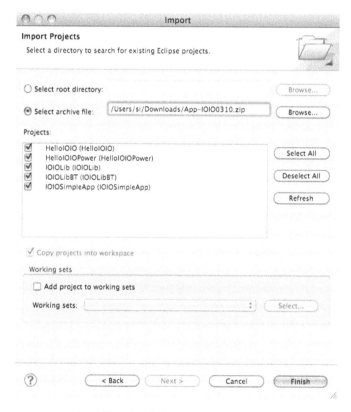

Figure 1-6. *Installing the IOIO library and samples in Eclipse*

Include all the example projects, as although you only actually need HelloIOIO, IOIOLib and IOIOLibBT,→ it is useful to have the other projects as examples. The projects that you find here may vary as the IOIO software is updated.

You will then have the entries shown in Figure 1-7 in your Project Explorer. If there is a red error cross next to any of the projects, go to the Project menu and select Clean to clean all the projects. This will normally remove any crosses. If it isn't, try pasting the error message that you get into your search engine.

Figure 1-7. *Sample IOIO apps in the Project Explorer*

Hello IOIO

Open the first project in the list (HelloIOIO) and connect your phone to your computer with the USB lead. For the app to be installed onto your phone, you will need to make sure that USB debugging is turned on. You can find this option on your phone if you open Settings and then go to Applications and then Development (Figure 1-8).

Select HelloIOIO in the Eclipse Project Explorer and then click on the green Play button on the toolbar. The first time you do this, you may get the dialog shown in Figure 1-9. Just select the option *Android Application*.

If your phone is connected correctly, the App will be installed and launched on it, so that you can now disconnect the computer end of the USB lead and plug it into the IOIO board, which should also be connected to your power supply (Figure 1-10).

Clicking on the button will turn the LED on the IOIO board on and off.

It is beyond the scope of this book to teach you Android and Java programming from scratch. However, the IOIO library is very nicely designed, and you should find that even if you have very little programming experience, you will be able to take the programs in this book and modify them for your own use.

Figure 1-8. *Turning on USB debugging*

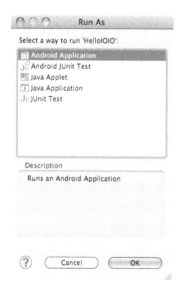

Figure 1-9. *Application Type dialog*

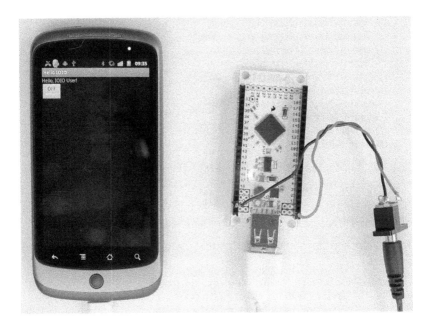

Figure 1-10. *Hello IOIO*

In the HelloIOIO app, there are really just two files that do most of the work: *MainActivity.java* and *main.xml*.

MainActivity.java can be found under *src/ioio/example/hello*. This file contains the code that controls the one and only Activity in this project. In Android, an Activity is akin to a screen in other frameworks.

If you look at the top of this file, you will find:

```
public class MainActivity extends AbstractIOIOActivity {
```

This tells us that we are creating a subclass of AbstractIOIOActivity, which will implement the IOIO framework behind the scenes for us, so we can just get on with the part of our project that is specific to this app.

This Java file has no information about how the user interface for this Activity is arranged. This is held in the template file *main.xml*, which is kept in *res/layout*:

```
<?xml version="1.0" encoding="utf-8"?>
<LinearLayout xmlns:android="http://schemas.android.com/apk/res/android"
    android:orientation="vertical"
    android:layout_width="fill_parent"
    android:layout_height="fill_parent"
    >
```

```
<TextView
    android:layout_width="fill_parent"
    android:layout_height="wrap_content"
    android:text="@string/hello"
    android:id="@+id/title"/>
<ToggleButton android:text="ToggleButton"
android:layout_width="wrap_content"
  android:layout_height="wrap_content"
  android:id="@+id/button">
</ToggleButton>
</LinearLayout>
```

This XML file contains a LinearLayout tag, which in turn contains TextView and ToggleButton tags.

Note that both the user interface controls have an **android:id** attribute that will be used in **MainActivity** to obtain a handle on them.

Turning back to *MainActivity.java*, the first method we come across is called onCreate.

```
public void onCreate(Bundle savedInstanceState) {
        super.onCreate(savedInstanceState);
        setContentView(R.layout.main);
        button_ = (ToggleButton) findViewById(R.id.button);
}
```

This will be called when the Activity is created. After invoking onCreate on the superclass, it associates our layout with the Activity and then creates a link to a member variable that holds a reference to the toggle button.

Android is very fussy about what happens on its UI thread. In fact, you cannot do anything here that might block the UI thread, even for a second or two. If you do, Android will decide that the app is not responding and throw a fatal exception. For this reason, we put all the IOIO processing activity into a separate class that subclasses AbstractIOIOActivity.IOIOThread:

```
class IOIOThread extends AbstractIOIOActivity.IOIOThread {
        private DigitalOutput led_;

        @Override
        protected void setup() throws ConnectionLostException {
                led_ = ioio_.openDigitalOutput(0, true);
        }

        @Override
        protected void loop() throws ConnectionLostException {
                led_.write(!button_.isChecked());
                try {
                        sleep(100);
                } catch (InterruptedException e) {
                }
```

```
        }
    }
```

This class requires two methods to be implemented: **setup** will be called whenever the IOIO detects that the phone has been connected, whereas **loop** is called repeatedly and indefinitely.

We have a member variable called **led_** that is an instance of **DigitalOutput**. The class **DigitalOutput** is responsible for implementing functionality concerned with using a GPIO pin as a digital output. This amounts to setting the output to high or low using the **write** method.

The **setup** method simply sets up the instance of **DigitalOutput** held in the variable **led_**. The first argument to **openDigitalOutput** is the pin to use—in this case, pin 0. Looking closely at the IOIO board, you will see that there is no pin 0. Pin 0 is actually reserved for the onboard LED. Later we will change this example to use an external LED and change the pin number to 46.

The second argument to **openDigitalOutput** is the initial state of the pin, which in this case is **true**, meaning **high** or 3.3V.

Looking at the **loop** method, we can see that all that happens is that we use **button_isChecked()** to determine the state of the toggle button, and set the output of the LED pin to be the inverse of that.

We then have a try/catch construction around a call to **sleep**. Your loop should include a sleep to allow this thread to yield and allow other threads to have a chance to do something. Any exception from **sleep** will just be ignored.

Back in the **MainActivity** class itself, we have the following glue code, which will be present in any project and creates the IOIO thread:

```
@Override
protected AbstractIOIOActivity.IOIOThread createIOIOThread() {
        return new IOIOThread();
}
```

Connecting Things to IOIO

So now, we can turn on an LED on the IOIO from our phone, but in the projects that follow, we are going to be connecting external components to the board. That is, after all, the purpose of an interface board. We certainly have plenty of sockets into which we can plug things, but before we do that, we need to know a little more about those connections.

Looking back at Figure 1-1, you can make out the labels printed next to each connection. At the end of the board nearest the USB connector, we have these power connections:

VIN

   The supply voltage between 5V and 15V DC. This is best thought of as
   the input voltage to the board.

3.3V

   A 3.3V regulated supply from a voltage regulator IC on the IOIO.

5V

   A 5V regulated supply from a voltage regulator IC on the IOIO

GND

   Ground or 0V

The IOIO board is primarily a 3.3V board. That is, all the inputs and outputs
are designed to work at 3.3V. However, many electronic modules and devices are designed to work at 5V. This is why the IOIO also provides a 5V supply
and some of its pins are capable of being pulled-up to 5V—but not all.

   WARNING: Incorrect application of 5V to a 3.3V
   connection could damage your IOIO.

The pins not associated with the power supply are just numbered sequen-
tially. These pins can all be used as a GPIO or General Purpose Input Output
pin. That is, when used as outputs, they can be set to 3.3V or 0V (GND), and
when used as digital inputs they can tell whether the voltage is above or
below a threshold voltage of about 1.5V. Many of these pins can also be used
as analog inputs, PWM outputs (a kind of analog output), and some as outputs that can tolerate 5V.

Some of the pins can be used for TWI (Two Wire Interface) communications
with certain peripherals. Other pins can be used for serial communication to
computers, Bluetooth modules, etc., using one of the four UARTs (Universal
Asynchronous Receiver Transmitters).

Table 1-2 summarizes the features available.

Table 1-2. *IOIO pins*

Usage	Pins
Analog in	31-34, 37-46
TWI (data, clock)	(4, 5), (26, 25), (47, 48)
UART	3-7, 9-14, 27-32, 34-40, 45-48
5V-friendly	3-7, 10-14, 18-26, 47-48

Just to prove that we can attach some external electronics to our IOIO, we will attach an LED to one of the pins and then modify the Hello IOIO example to use this LED rather than the LED built onto the IOIO board itself.

To do this, you will need an LED and a resistor. Just about any LED between 100Ω and 270Ω will do fine. The parts are listed in Table 1-3.

Table 1-3. *Parts bin*

Quantity	Description	SparkFun SKU	Farnell code
1	Red 5mm LED	COM-09590	1712786
1	100Ω 0.5W metal film resistor		9340300

The longer lead of the LED is the positive lead. This is be inserted into the socket for pin 46 on the IOIO. Bend out the other lead of the LED and twist it together with one lead of the resistor. Push the unconnected lead of the resistor into one of the GND sockets, as shown in Figure 1-11.

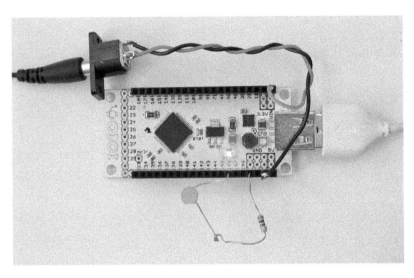

Figure 1-11. *Attaching an LED to IOIO*

All that remains is to reconnect our phone to our computer so that we can change the LED pin to 46.

Open the file *MainActivity.java* in Eclipse and change the line:

```
led_ = ioio_.openDigitalOutput(0, true);
```

to read:

```
led_ = ioio_.openDigitalOutput(46, true);
```

Redeploy the application to your phone and then plug your phone into the powered up IOIO. You should now be able to turn the external LED on and off.

Conclusion

In the projects that follow, we will use the IOIO in various different ways. We will make use of digital and analog inputs and outputs to build a series of projects.

Take some time to select a project that you would like to build, order the parts and then have some fun making it. Photographs, videos, source code, and pre-built apps for the project can all be found at the website for the book.

2/Intruder Alarm

In this chapter we'll develop a motion sensor intruder alarm with a difference: it sends you a text message when it detects movement. The project has very few components apart from the IOIO. It uses just a PIR movement detector and a resistor.

Figure 2-1 shows the interface to the IOIO Alarm. There are two fields, one for the phone number to send the text to and one for the message to be sent when movement is detected.

Under those fields appear two toggle buttons. The first toggles test mode on and off. When in test mode, no actual text messages are sent. A message appears on the screen momentarily instead. The second toggle button is labeled RUN; pressing this will start the monitoring after a period of 10 seconds, which gives you enough time to leave the room.

 WARNING: This project sends SMS text messages, which may cost you money. So do not get carried away when you are testing it.

Hardware

In addition to your IOIO, which you should have kitted out with sockets and a power lead, you will need to buy the items in Table 2-1.

Table 2-1. *Parts bin*

Quantity	Description	SparkFun SKU	Farnell code	Newark code
1	PIR Module	SEN-08630		
1	10 kΩ 0.5W metal film resistor		9339787	38K5141

The schematic diagram for the project is shown in Figure 2-2.

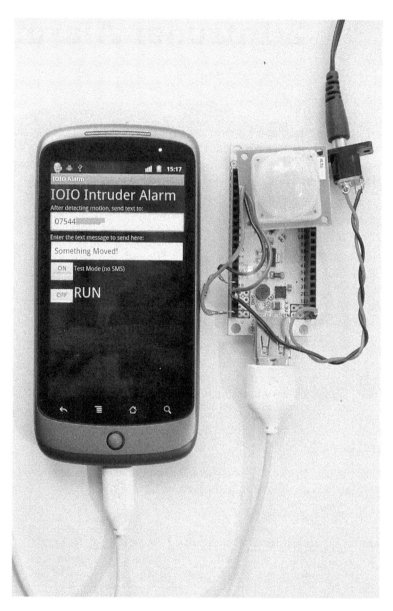

Figure 2-1. *IOIO intruder alarm*

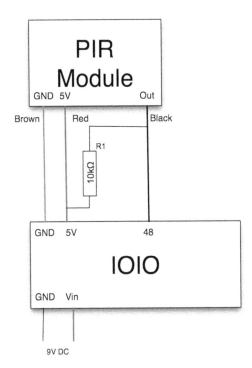

Figure 2-2. *The schematic diagram*

The PIR (Passive Infra Red) module detects movement. When something in front of its field of view moves, it turns a transistor on. The output is of a type called "open collector" and requires a pull-up resistor of 10 kΩ to be connected between its output and +5V. The output will actually normally be at 5V, but will fall to 0V when movement is detected. For this reason, we must use one of the IOIO's 5V tolerant pins (in this case, pin 48).

You might expect the black lead from the PIR module to be GND, but it's actually the output and the brown lead is GND.

Since there are so few connections to make, we are just going to push leads into the sockets. This is not the most reliable way of connecting leads into the sockets, but if you thicken up the leads with a layer of solder and put a little kink in them, a pretty good connection can be made.

The first step in the construction is to cut off the connector socket of the PIR module and strip and tin the ends of the leads. Then solder the red and black leads to either lead of the resistor, as shown in Figure 2-3.

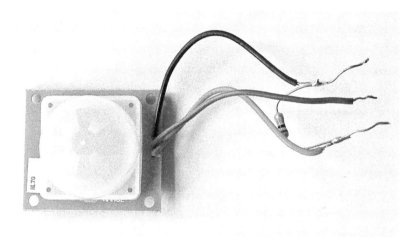

Figure 2-3. *Preparing the PIR module (soldering leads to the resistor)*

The red-lead side of the resistor will go in the +5V socket, the black in the socket for pin 48 and the brown in one of the GND sockets as shown in Figure 2-4, where you can also see the power lead that we made in Chapter 1.

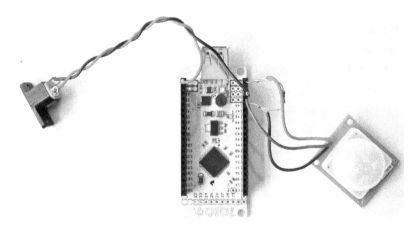

Figure 2-4. *Preparing the PIR module (attaching resistor leads)*

That is all there is to the hardware. Now we need to turn our attention to the software side.

Software

The source code for the app can also be downloaded from the book's website.

Much of the framework for the app is similar to that of the HelloIOIO example of Chapter 1, so we will just look at the parts of the code concerned with interfacing to the PIR sensor.

```
protected void setup() throws ConnectionLostException {
        pir_ = ioio_.openDigitalInput(48, Mode.FLOATING);
        led_ = ioio_.openDigitalOutput(0);
}
```

The **setup** method opens a digital input on pin 48 for the PIR sensor and a digital output for the built-in LED on the IOIO. This LED will flicker on whenever movement has occurred.

When opening a digital Input using **openDigitalInput**, you supply two arguments. The first argument is the pin to open, and the second specifies the mode. This mode can be one of **Mode.FLOATING**, **Mode.PULL_UP** or **Mode.PULL_DOWN**. We have used **FLOATING**, because the built-in resistors that can be assigned with **PULL_UP** and **PULL_DOWN** are too weak for the PIR sensor and we need to use an external pull-up resistor:

```
protected void loop() throws ConnectionLostException {
        boolean wasMovement = false;
        try {
                wasMovement = ! pir_.read();;
        } catch (InterruptedException e1) {
                e1.printStackTrace();
        }
        led_.write(! wasMovement); // LED false = on
        if (wasMovement) {
                movementCount ++;
        }
        long now = System.currentTimeMillis();
        if (now > lastTime + 1000) {
                // every second
                lastTime = now;
                if (movementCount > 50) {
                    if (now > startTime_ + 10000 && runButton_.isChecked()) {
                                handleAlarm();
                        }
                }
                movementCount = 0;
        }
        try {
                sleep(10);
        } catch (InterruptedException e) {
```

```
        }
    }
```

The built-in LED is confusingly wired in such a way that when its output is set to false, it turns on, and when it is set to true, it turns off.

The PIR sensor is quite sensitive and we do not want too many false positives. So, rather than trigger an alarm every time the PIR sensor detects movement, we will consider it to be an alarm only if there are more than 50 such events in the 100 times that we check per second.

To do this, we use two member variables: `lastTime` and `movementCount`. The variable `lastTime` is a long integer and is assigned to the last time that the count was checked. We can use a long integer to represent time, because the `System.currentTimeMillis()` call returns us a system time in milliseconds. The following line determines whether a whole second has passed since the last time we checked:

```
if (now > lastTime + 1000) {
```

If it has been at least a second, this is where we test to see if both 10 seconds have elapsed since the Run button was pressed (using another member variable, `startTime_`) and that the Run button was checked. If all of this is true, we call `handleAlarm`:

```
private void handleAlarm() {
        if (testButton_.isChecked()) {
                toast("Test Mode, no SMS sent");
        }
        else {
                sendSMS();
        }
        runOnUiThread(new Runnable() {
                @Override
                public void run() {
                        runButton_.setChecked(false);
                }
        });
}
```

The `handleAlarm` method decides whether to send a real text message or just make a "toast" notification, depending on the state of `testButton`. Once an alarm has been triggered, the Run button is unchecked to prevent further triggering and text messaging.

The code to uncheck the button has to be run using the `runOnUiThread` command.

One refinement of the app is that it remembers the phone number and message to send, even after the app quits. It does this using the Android preferences mechanism. In the onCreate method, as well as the usual assignment of user interface controls to remember variables, we also set their default values:

```
public void onCreate(Bundle savedInstanceState) {
    super.onCreate(savedInstanceState);
    setContentView(R.layout.main);
    SharedPreferences settings = getSharedPreferences(PREFS_NAME, 0);
    sms_ = (TextView)findViewById(R.id.sms);
    message_ = (TextView)findViewById(R.id.message);
    testButton_ = (ToggleButton)findViewById(R.id.testButton);
    testButton_.setChecked(true);
    runButton_ = (ToggleButton)findViewById(R.id.runButton);
    sms_.setText(settings.getString("sms", ""));
    message_.setText(settings.getString("message", "Something Moved!"));
    runButton_.setOnCheckedChangeListener(this);
}
```

The following line retrieves a settings object for the app:

```
SharedPreferences settings = getSharedPreferences(PREFS_NAME, 0);
```

Then individual setting values can be retrieved using the following syntax, where the first argument is the name of the setting and the second is a default value if there is no value found:

```
message_.setText(settings.getString("message", "Something Moved!"));
```

The settings are actually saved whenever one of the toggle buttons changes state in the onCheckedChanged handler:

```
@Override
public void onCheckedChanged(CompoundButton buttonView, boolean isChecked) {
        if (isChecked) {
                toast("You have 10 seconds before sensing starts");
                startTime_ = System.currentTimeMillis();
        }
        // save the fields in prefs so they are there next time
    SharedPreferences settings = getSharedPreferences(PREFS_NAME, 0);
    SharedPreferences.Editor editor = settings.edit();
    editor.putString("sms", sms_.getText().toString());
    editor.putString("message", message_.getText().toString());
    editor.commit();
}
```

This method is also responsible for setting a time stamp in the member variable startTime_ to delay activation of the alarm.

Conclusion

This is a useful little alarm. One slight flaw is that there is no way to deactivate the alarm without triggering it, so a text message will get sent every time you use it. The app could be modified to provide a delay after triggering, during which the alarm could be disabled (perhaps using a secret code). Another refinement could be to allow the app to capture a photo with the phone's onboard camera and send this as part of the text message.

In the next chapter, we will look at a very simple project that just uses a temperature sensor to record temperatures and save them on the phone's micro SD card.

3/Bluetooth Temperature Logger

This project (Figure 3-1) is probably the simplest of the projects in this book. It is very easy to make and there is no soldering to do, other than what you did in Chapter 1 to prepare your IOIO by adding sockets to it.

The temperature sensor itself is a small, 3-pin chip that just plugs into the sockets on the IOIO.

The controlling software takes a temperature reading every 10 seconds and logs it onto the SD card in your phone. It also displays the current temperature in degrees Centigrade or Fahrenheit. When you want to retrieve the data that has been collected, you can just use the USB storage feature of your phone to transfer the file onto your computer. The data is stored in a CSV format so that it can be imported into a spreadsheet.

What's more, we are going to use a USB Bluetooth module attached to the USB port of the IOIO to communicate with your phone wirelessly.

If you don't want to use Bluetooth for this project, you can just plug the phone into the IOIO in the same way as the project in Chapter 1.

Hardware

In addition to your IOIO, which you should have kitted out with sockets and a power lead in Chapter 1, you will need to buy the items in Table 3-1.

Table 3-1. *Parts bin*

Quantity	Description	SparkFun SKU	Farnell code	Newark code
1	TMP36	SEN-10988	1438760	19M9015
1	USB Bluetooth adaptor		1848138	39T4089

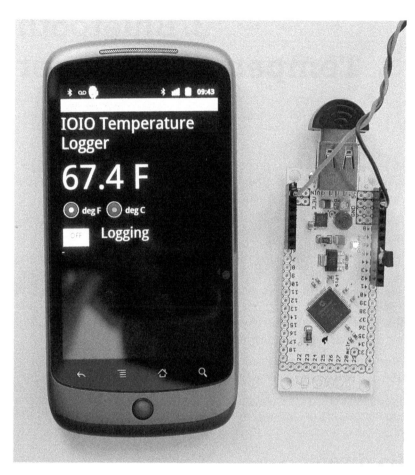

Figure 3-1. *IOIO temperature logger*

Almost any USB 2.0 Bluetooth adaptor should work with this project. These can be bought for as little as 2 USD. The IOIO Over Bluetooth page lists some devices that have been tested and are known to work: [*https://github.com/ytai/ioio/wiki/IOIO-Over-Bluetooth*]

The schematic diagram for the project is shown in Figure 3-2.

The TMP36 sensor uses only a tiny amount of current and so, to plug it directly into the IOIO, we can use two GPIO sockets to provide the +3.3V and GND connections that it needs. Its output will be connected to pin 45, which is used as an analog input.

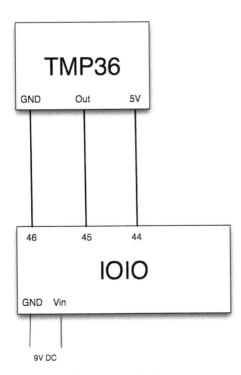

Figure 3-2. *The schematic diagram*

The voltage at this input will be proportional to the temperature, and so the analog reading can be converted into a temperature with a bit of math.

Putting a slight kink in the leads will ensure a good connection with the socket. Figure 3-3 shows the temperature sensor fitted into sockets 44 to 46. Make sure you get the device the right way around, with the curved side towards the center of the board.

The USB Bluetooth adaptor is just pushed into the USB socket.

Setup

You can download the app for this project from the book's website, where you will also find a link to the source code. Install the app, and connect the power to your IOIO board.

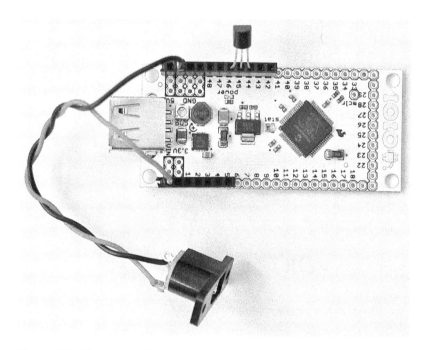

Figure 3-3. *The temperature sensor*

For the Bluetooth link to work, you need to pair the Bluetooth adaptor in the IOIO with the phone. To do this, open the Settings on your phone, and select "Wireless and Networks" then "Bluetooth Settings". This should start your phone scanning for devices, after which you should see a list that includes IOIO (Figure 3-4).

Select the IOIO device from the list and you will be prompted to pair with the device (Figure 3-5). Enter the PIN 4545.

Now, when you start the app, you should see it displaying the current temperature. If you click on the *Logging* button, then every 10 seconds, the temperature will be written to a file on the phone's SD card. The file name will be *temp*, followed by the date. The file is written in CSV format so that you can open it directly with most Spreadsheet software.

Software

Much of the software in this project is very similar to the Intruder Alarm project, so just a few things are highlighted here.

Figure 3-4. *Searching for the IOIO on Bluetooth*

Firstly, there is the USB functionality. You might go looking in the code for it, but it isn't there! The IOIO libraries are written in such a way that there is literally no code to write to make the app work with Bluetooth. This means that we could do any of the projects in this book with Bluetooth instead of USB, without changing a line of code. Well, this is not quite true, as the project in the next chapter relies on being able to turn the pins on and off quickly, and Bluetooth is not quite fast enough.

Figure 3-5. *Pairing with the IOIO on Bluetooth*

Code worth highlighting in this project is that used to write to the SD card. The method that does this is called `appendToFile`:

```
private void appendToFile(String filename, String line) {
    File root = Environment.getExternalStorageDirectory();
    try {
        FileOutputStream f = new FileOutputStream(new File(root, filename),
true);
        f.write(line.getBytes());
        f.close();
```

```
    } catch (Exception e) {
      toast(e.getMessage());
    }
  }
```

This opens the file in append mode (creating it if it doesn't exist) and then adds a line formatted with the time, the temperature reading, and the units (F or C).

For this to work, the following permission has to be added to the project manifest file, *AndroidManifest.xml*:

```
    <uses-permission
  android:name="android.permission.WRITE_EXTERNAL_STORAGE">
    </uses-permission>
```

Conclusion

This is a nice easy project, but one that could be extended into something more sophisticated. Other sensors could be added and you could use the Android phone to forward on the readings to a web service such as Pachube.

In the next project, we get a lot more complex and have to do some serious construction work for a light show project.

4/LED Matrix
Light Show

This chapter uses a multicolor LED matrix to make a fun charger for your Android phone (Figure 4-1). In this particular case, "fun" takes the form of a marching Space Invader animation—and if that isn't fun, I don't know what is!

The project has a variety of different modes: it can just display a static image, or it can display an animation, or it can make use of the phone's microphone to provide a spectrum type display. If you start your music player on the phone before you start the IOIO Matrix app, it will respond to the sounds coming from your phone.

This is the only project in the book that will not work using a Bluetooth adaptor rather than a USB cable. This is because Bluetooth simply isn't fast enough to send the commands to the pins to refresh the display.

The Design

In this design, the IOIO board uses 24 of its pins to control the LED Matrix. It requires 24 pins because the LED Matrix is arranged as a grid of LEDs. Each cell in the matrix actually has two LEDs in it, one red and one green. This is used to set the color of any individual cell to red, green or—if both LEDs are lit at the same time—orange.

Schematic

Figure 4-2 shows the schematic diagram for the project.

The anodes of the LEDs are each driven by a GPIO pin on the IOIO. Each has a series resistor to limit the current to the LED.

Because 8 LEDs share each common cathode connection to ground, there would be too much current flowing for a GPIO pin to sink, so a MOSFET transistor is used to switch each column in turn. The gate of each MOSFET is connected directly to a GPIO pin.

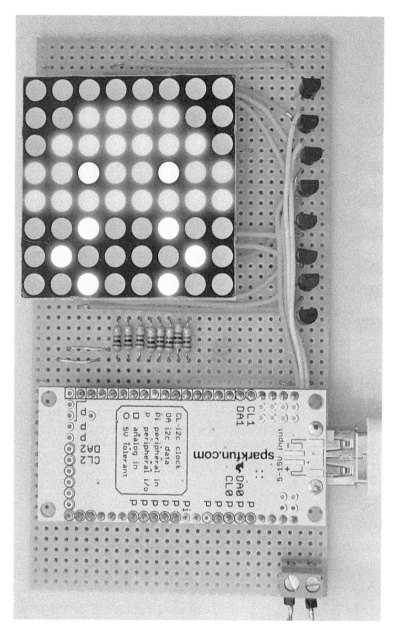

Figure 4-1. *LED matrix light show*

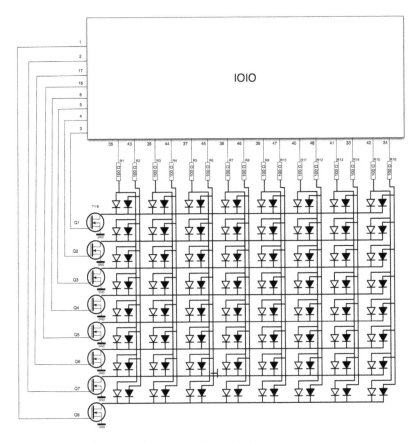

Figure 4-2. *Schematic diagram for the project*

Wiring Diagram

The project is built on a piece of stripboard. Stripboard is a kind of proto-typing board, with parallel tracks of copper running on one side of the board. Component leads are pushed through from the top and soldered to the copper track below.

On one side of the stripboard is a set of header pins designed to accept the IOIO board with its header sockets attached. The IOIO board will be fitted upside down onto the headers. The other side contains header sockets into which the LED Matrix is fitted.

A screw terminal for Vin and GND is used to simplify the process of providing power to the project.

Figure 4-3 shows the stripboard layout for the project.

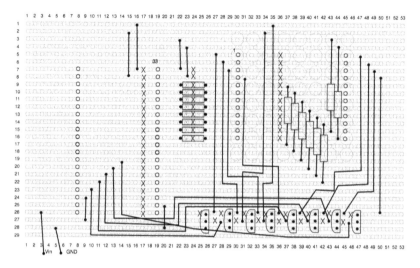

Figure 4-3. *The stripboard layout*

Construction

You will need the following parts to construct this project.

Table 4-1. *Parts bin*

Quantity	Description	SparkFun SKU	Farnell code	Newark code
1	8 x 8 Dual-color LED Matrix	COM-00682		
8	2N7000 MOSFETS		9845178	89K1814
16	100Ω 0.5W metal film resistor		9339760	58K3723
2	SIL Header socket strip	PRT-00115	1217038	52K3454
2	SIL header pins	PRT-00116	1097954	93K5128
1	Screw terminal block	PRT-08084	1641932	19P1412

Quantity	Description	SparkFun SKU	Farnell code	Newark code
1	Stripboard 29 strips each of 53 holes		1201473	96K6336

Step 1. Prepare the Stripboard

The first step is to cut the stripboard to the correct size. The best way to do this is to use a craft knife to heavily score a line through the holes on the line below the last strip or column you need, and then break the board over the edge of your work desk. Be careful doing this, as it can leave sharp edges.

You then need to break the track in the positions indicated by an X in Figure 4-3. I find it useful to mark rows and columns 10, 20, 30, etc., on the top of the board to find the right position for the break and then push a wire through to find the position on the track side of the board. I use a drill bit, twisted between my fingers to remove the copper.

Figure 4-4 shows the copper side of the board, with all the breaks drilled.

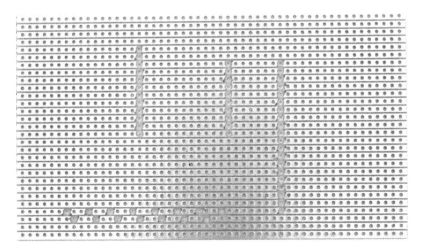

Figure 4-4. *Breaks drilled in the stripboard*

You may find it easier to work from Figure 4-4 than Figure 4-3. When you have made all the breaks, go back and inspect every one carefully to make sure that there is no trace of copper remaining, as this could cause a short and potentially destroy your IOIO. If in doubt, you can also use your multimeter on continuity mode, to make sure the break is clean.

Step 2. Fit the Link Wires

The copper tracks on the bottom will anchor our components and make some of the connections. However, there are a lot more connections to be made with linking wires. The longer leads should be made using insulated solid core wire, and the shorter connections can just be bare wire.

Using Figure 4-3 as a reference, solder link wires into place. Note that this is not a quick job. You should put aside an hour to do this, because there are a lot of links to put in place.

Do not be tempted to solder the header pins in place first. Although this would make it much easier to work out where the link wires need to go, it makes it much harder to solder the links themselves into place, as they will just fall out when you turn the board upside-down to solder it.

When all the links are in place, you should have a board that looks like Figure 4-5.

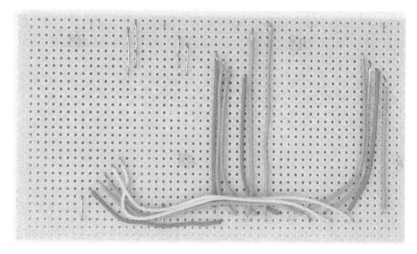

Figure 4-5. *The stripboard with resistors in place*

Step 3. Fit the Resistors

The next step is to fit the next lowest parts, which are the resistors. Again, using Figure 4-3 as a reference, solder them into place. When all the resistors are in position, your board should look like Figure 4-6.

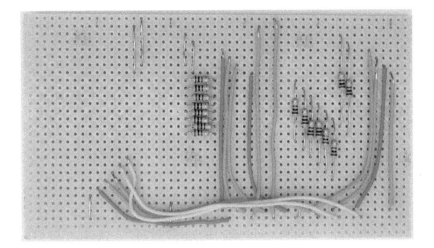

Figure 4-6. *The stripboard with links in place*

Step 4. Fit the Headers

You can make it easier to fit the header plugs into the IOIO and the sockets into the LED Matrix by fitting the components into the headers before putting them in the right position on the board. Double-check that the placement is correct, as it will be hard to unsolder them once they are in position.

If the header strips are not the right lengths, you will need to cut them to the right number of connections using a craft knife. When cutting the sockets, this will usually mean sacrificing one of the socket connections, so cut through the socket after the number you need, rather than try and cut between sockets.

Once the sockets are in place, your board should look like Figure 4-7.

Step 5. Fit the MOSFETs

The last components to be added to the board are the MOSFET transistors. Be careful to ensure that they are the right way around, and solder them into place, raised about 1/4 inch above the surface of the board.

Figure 4-8 shows the board with the MOSFETs in place.

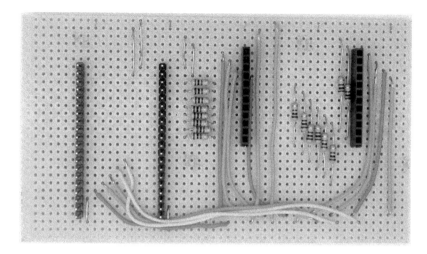

Figure 4-7. *The stripboard with headers in place*

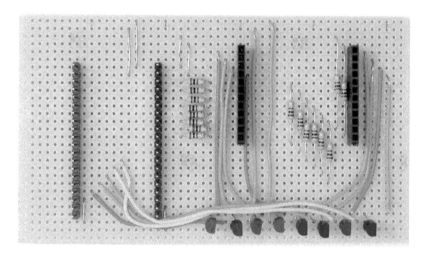

Figure 4-8. *The stripboard with MOSFETs in place*

Step 6. Fit the Power Terminal and IOIO

That's pretty much all the hardware. It just remains to solder the screw terminal for the power into place, and then fit the LED Matrix and IOIO (Figure 4-9).

Solder the screw terminal block into place first and mark the upper connection with a + to reduce the chance of applying the supply voltage reversed.

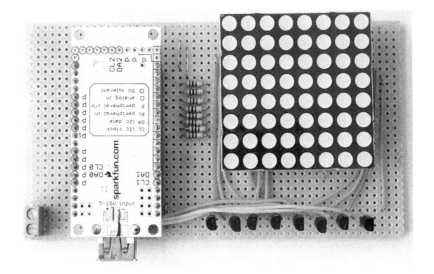

Figure 4-9. *The stripboard fully assembled*

The LED Matrix has little cutouts and pegs to allow bigger displays to be made by joining more than one of them together. The correct orientation for the cutouts is at the bottom and left of the board.

Figure 4-9 shows the board fully assembled and ready to go.

Software

They key to this project is being able to refresh the screen fast enough. This is all wrapped up in the chain of methods in *MainActivity.java* that start with `refreshMatrix`:

```
private void refreshMatrix() throws ConnectionLostException {
  for (int col = 0; col < 8; col++) {
    clearPreviousColumn(col);
    displayColumn(col);
    delay(3);
  }
}
```

This iterates for each column, clearing its previous settings, displaying the new column, and then sleeping for 3 milliseconds:

```
private void clearPreviousColumn(int col) throws ConnectionLostException {
  int columnToClear = col - 1;
  if (columnToClear == -1)
  {
```

```
      columnToClear = 7;
    }
    cc[columnToClear].write(false);
    for (int row = 0; row < 8; row++) {
      r[row].write(false);
      g[row].write(false);
    }
  }
```

Clearing the previous column is a matter of finding the column before the current one, including wrapping round. We then turn off the common cathode on the LED matrix for the column in question and then turn off all the red and green anodes for all 8 rows:

```
private void displayColumn(int col) throws ConnectionLostException {
  cc[col].write(true);
  for (int row = 0; row < 8; row++) {
    r[row].write((display_[col][row] & 1) > 0);
    g[row].write((display_[col][row] & 2) > 0);
  }
}
```

Displaying the new column involves turning on the appropriate common cathode of the LED matrix, and then setting the red and green anodes according to the current column of the 2D array of colors held in the member variable **display_**, which looks something like this:

```
private int[][] testPattern1_ = {
    {1,1,1,1,1,1,1,1},
    {1,2,2,2,2,2,2,2},
    {1,2,3,3,3,3,3,3},
    {1,2,3,1,1,1,1,1},
    {1,2,3,1,2,2,2,2},
    {1,2,3,1,2,3,3,3},
    {1,2,3,1,2,3,1,1},
    {1,2,3,1,2,3,1,2}
};
```

The number 0 means both LEDs are off; 1, red LED; 2, green LED; and 3, both LEDs on (orange).

Everything else in this app, including the animation, is just a matter of assigning **display_** to a different 2D array every half second.

The frames of the animation are defined in a separate class file, which is also responsible for providing a value (**frameDelay**) for the time between frames in milliseconds:

```
package com.ioiobook.matrix;

public class TestAnimation {
```

```
        public final static int frameDelay = 500;

        public final static int[][][] animation = {
            { //1
                    {0,0,1,1,1,1,0,0},
                    {0,1,1,1,1,1,1,0},
                    {1,1,2,1,1,2,1,1},
                    {1,1,1,1,1,1,1,1},
                    {0,0,3,0,0,3,0,0},
                    {0,0,3,0,0,3,0,0},
                    {0,0,3,0,0,3,0,0},
                    {0,0,0,0,0,0,0,0}
            },
            { //2
                    {0,0,0,0,0,0,0,0},
                    {0,0,1,1,1,1,0,0},
                    {0,1,1,1,1,1,1,0},
                    {1,1,2,1,1,2,1,1},
                    {1,1,1,1,1,1,1,1},
                    {0,0,3,0,0,3,0,0},
                    {0,3,0,0,0,0,3,0},
                    {0,0,3,0,0,3,0,0}
            },
```

The spectrum display makes use of a third-party open source library, wrapped up in a class (SpectrumDrawer.java). This is instantiated with a display to draw on:

```
public class SpectrumDrawer {

    private float gain_ = 1000000.0f;
    private int[][] displayArray_;
    private Window win_;
    private FFTTransformer spectrumAnalyser_;
    private int historyIndex_;
    private float[] average_;
    private float[][] histories_;

    // 128 values in average_ we just want 8 - Fn = n * Fs / N
    // where Fn is freq at data point n, Fs is the sample freq
    // and N is the buffer size
    private final int[] frequencies_ = { 2, 4, 6, 10, 15, 25, 55, 80 };
    private final int[] colors_ = { 2, 2, 3, 3, 3, 1, 1, 1 };

    public SpectrumDrawer(int[][] display) {
        displayArray_ = display;
        win_ = new Window(MainActivity.AUDIO_BUFFER_SIZE,
                Window.Function.BLACKMAN_HARRIS);
        spectrumAnalyser_ = new
FFTTransformer(MainActivity.AUDIO_BUFFER_SIZE, win_);
```

```
            average_ = new float[MainActivity.AUDIO_BUFFER_SIZE / 2];
            histories_ = new float[MainActivity.AUDIO_BUFFER_SIZE / 2]
    [MainActivity.AUDIO_BUFFER_SIZE / 2];
        }

        public void calculateSpectrum(short[] buffer) {
            // apply FFT to the buffer to get the spectrum,
            // but we only have 8 columns
            // so sum into 8 bands
            spectrumAnalyser_.setInput(buffer, 0,
    MainActivity.AUDIO_BUFFER_SIZE);
            spectrumAnalyser_.transform();
            historyIndex_ = spectrumAnalyser_.getResults(average_, histories_,
                    historyIndex_);

            for (int c = 0; c < 8; c++) {
                int resultIndex = frequencies_[c];
                // Do we need to log this?
                int power = (int) (Math.log(average_[resultIndex] * gain_));
                Log.d("SRM", "" + power);
                if (power > 7)
                    power = 7;
                for (int r = 0; r < 8; r++) {
                    if (power > r) {
                        displayArray_[7 - r][c] = colors_[r];
                    } else {
                        displayArray_[7 - r][c] = 0;
                    }
                }
            }
        }
    }
}
```

When the `calculateSpectrum` method is called, a Fast Fourier Transform
(FFT) is applied to a sample of the audio from the phone's microphone.

A FFT is used in this case to take a sample of an audio file and find the relative
sizes of each of the frequencies that make up the sound. This produces an
array of the power of a range of frequencies. We can then pick off frequencies
from this and use them to set the colors of the matrix display.

The `histories_` array is required by the third-party library to provide aver-
aging of the FFT results.

For each column, we light a number of LEDs in the rows equal to the power.
The actual color of each of the lit LEDs is determined by the `colors_` array.

To feed the `SpectrumDrawer` with new data, a separate thread is started in the
`onCreate` method of the `MainActivity` class:

```
AudioReader.Listener listener = new AudioReader.Listener()
{

        @Override
        public void onReadComplete(short[] buffer) {
                spectrumDrawer_.calculateSpectrum(buffer);
        }

        @Override
        public void onReadError(int error) {

        }

};
audioReader_ = new AudioReader();
audioReader_.startReader(F, AUDIO_BUFFER_SIZE, listener);
```

The thread is encapsulated in the **AudioReader** class. This class is provided in the **org.hermit** library.

Conclusion

This is quite a challenging project, both for the builder and the IOIO that has to keep updating the outputs to keep the display alive.

The app is intended as a starting point for your own experiments. There are many ways that it could be improved, including a file format for the animations to that they can be loaded and a optimization of the display mechanism to reduce flicker.

In the final chapter in this book, we are going to get more physical and make a little Bluetooth-controlled rover.

5/Surveillance Rover

The last project of this book is to create a small IOIO-powered rover. This is another project that uses a Bluetooth to give wireless control of the rover (Figure 5-1 and Figure 5-2).

Figure 5-1. *Surveillance rover*

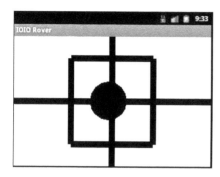

Figure 5-2. *Surveillance rover control software*

The rover also optionally provides a platform for a wireless web cam or a second Android phone with wireless web cam software on it, as shown in Figure 5-1. The author used an app called *IP Camera* from the Android Market, which converts the phone into a web cam that acts as a server over WiFi. You can then go to a URL in your browser and see the image from the web cam.

The Design

The outputs of an IOIO board are not powerful enough to drive electric motors, so a motor control breakout board is used. This little board (Figure 5-3) allows bi-directional control of the motors. That is, you can control both the speed and direction of two separate motors.

Figure 5-4 shows the schematic diagram for the project. This time, for obvious reasons, we will use batteries rather than a power adaptor.

The IOIO pins for the motor control are selected so that we can use header pins to plug one side of the motor control board directly into the IOIO board, reducing the amount of wiring needed.

Construction

In addition to a IOIO prepared with header sockets as described in Chapter 1, you will need the parts listed in Table 5-1 to construct this project.

Figure 5-3. *SparkFun motor controller*

Table 5-1. *Parts bin*

Quantity	Description	SparkFun SKU	Farnell	Pololu	Newark
1	Motor Control Board	ROB-09457			
1	Second Android phone or web cam				
2	Gearmotor			1122	
1	USB Blue-tooth adaptor		1848138		39T4089
1	SPST toggle switch		1661841		22K8977
1	Battery Box		1650687		31C0585
4	rechargeable AAA cells	Local electronics store			
1	SIL Header socket strip	PRT-00115	1217038		52K3454
1	SIL header pins	PRT-00116	1097954		93K5128

Quantity	Description	SparkFun SKU	Farnell	Pololu	Newark
1	Plastic case, 145 x 80 x 30mm	Local electronics store			
2	Wheels to suit gearmotors	Local model store			
1	Castor	Local hardware store			

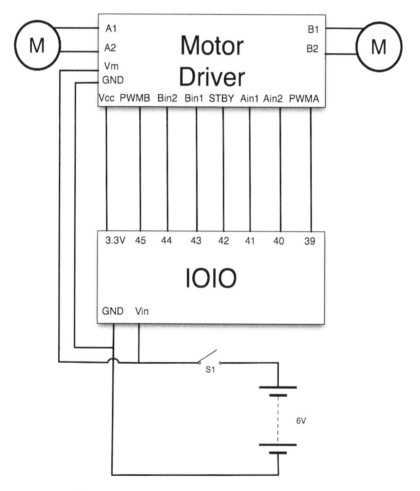

Figure 5-4. *Schematic diagram for the project*

Step 1. Prepare the Motor Controller

The motor controller has connectors on two sides. The control signals are all on one side, and we will attach a pin header to this side so that it can plug directly into the IOIO sockets. The other side will have a socket header attached to it so that we can wire the motors and supply to it. Figure 5-5 shows the motor controller with the pin header on one side and the sockets on the other.

Figure 5-5. *Preparing the motor controller*

Note that the two GND pins on the end of the motor connectors are not connected to the header socket, and header pins should each have 7 connections.

When the pins are in place, the motor controller will face inwards to the center of the IOIO and be plugged in to pins 39 to 45 (Figure 5-6).

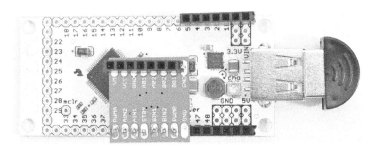

Figure 5-6. *The motor controller attached to the IOIO*

Step 2. Wire the IOIO to the Motor Controller

We need to make three power connections between the IOIO board and the motor controller (Figure 5-7). Break off a length of 7 header pins and fit it into the sockets on the motor shield and two lengths of two pins. Fit the first of these into the 3.3V and Vin sockets on the IOIO, and the other into two of the GND connections on the IOIO. We are going to solder the leads between these headers.

Figure 5-7. *Power connections between the IOIO and motor controller*

Solder the first wire between VM (motor voltage) on the motor controller and Vin on the IOIO. The second wire is between Vcc on the motor controller and 3.3V on the IOIO, and the final connection is between the GND pin on the motor controller and one of the GND connections.

Step 3. Prepare the Box Top and Motors

While this project is quite easy electronically, there is more mechanical construction than the other projects in this book. So when selecting a case, gearmotors, and wheels, make sure that everything will be able to fit easily in the box, and that the wheels fit the gearmotors and will be large enough to lift the whole box off the floor.

The box the author used was 145 x 80 x 30mm, which is quite a tight fit. Something slightly larger would be easier.

Figure 5-8 shows how the gearmotor's battery box and switch are laid out within the box.

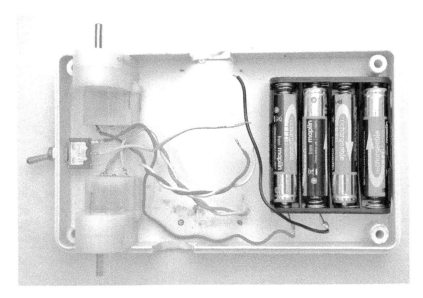

Figure 5-8. *The main components attached to the box*

Solder flying leads to both of the motors, the switch and the battery box. These must be long enough to easily reach the IOIO board that will be positioned in the center of the case. Use Figure 5-8 as a guide. The positive lead from the battery box is soldered to one side of the switch.

Drill holes in the box for the switch (and also for mounting the IOIO board), and then glue the gearmotors and battery box into place.

Depending on the size of your box, you may also need cut a hole for the Bluetooth adaptor, if there is not room for it to fit inside the enclosure (see Figure 5-9).

Step 4. Prepare the Box Base and Motors

The axles of the gearmotors are raised above the bottom half of the box, and so we need to cut the out a slot and hole for the axle, as shown in Figure 5-9.

Do not worry about the other holes in the box. The box was reused from another project.

The castor was attached to the front of the box using a hot glue gun. For a better idea of how the top and bottom of the box fit together, refer back to the finished project shown in Figure 5-1.

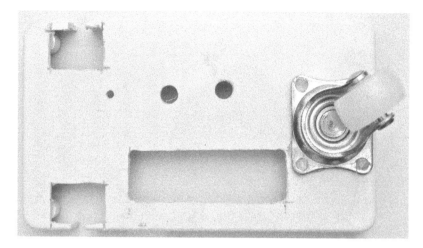

Figure 5-9. *Cutting the base of the box*

Step 5. Final Wiring

We can now attach the flying leads to the header pins in the IOIO and motor controller, as shown in Figure 5-10.

The connections to be made are listed below:

1. From the center connection of the switch to Vin on the IOIO. Note this pin will also have a connection going off to VM on the motor controller.

2. From the negative connection on the battery box to GND on the IOIO.

3. Both connections from one motor to AO1 and AO2 on the motor controller. Note that if these turn out to be the wrong way around, the motor will just turn in the opposite direction from the desired direction. If this happens, swap them over.

4. As above, but for the other motor to *BO1* and *BO2*.

Step 6. Testing

Before we fix the lid into place, we can test out the project with the rover on its back so that it doesn't go anywhere, but we can see what the motors are doing.

Insert the batteries and fit the Bluetooth adapter into the USB socket on the IOIO.

Load up the control app onto your phone from the book's website.

Figure 5-10. *Final wiring*

We are using a Bluetooth module, so this will need to be paired with your phone, as described in Chapter 3.

You should find that if you touch the dead center of the cross hairs, the motor will be off. Touching the north position should make both motors turn in a direction that would carry the rover forward. If this is not the case, then swap over the leads of the motor or motors that are not running in the right direction.

WARNING: Turn off the rover before doing this; an accidental short of the motor leads could damage the motor controller.

Once the rover is correct for moving forwards, touch the south position and the motors should spin the opposite way. The east and west positions should have the motors spinning in opposite directions.

If all is well, you can fix the two parts of the case together. But, before that, you may wish to make a mounting bracket for the second phone or web cam that is to be mounted onto the rover. If you use a web cam, you will have to figure out a power supply for it.

The author used a bit of plastic fixed to the same bolts that were used to mount the IOIO (Figure 5-1).

Software

There are quite a lot of pins used to control the motors (in fact, three for each motor). The *PWMA* and *PWMB* pins determine the speed of the motors. These use IOIO pins in PWM (Pulse Width Modulation) mode.

These pins are set up using the following method call:

```
pwma_ = ioio_.openPwmOutput(PWMA_PIN, PWM_FREQ);
```

The first argument is the pin to use, the second is the frequency of the pulses.

When it comes to actually setting the speed, we use the call below:

```
pwma_.setDutyCycle(Math.abs(left_));
```

The argument to the **setDutyCycle** method is a number between 0 and 1, where 0 is off and 1 is full speed.

The other pins used are all digital outputs that are either on or off. The pins *AIN1* and *AIN2* control the direction of the motor. If *AIN1* is high and *AIN2* is low, the motor will spin one way. If you reverse that so that *AIN1* is low and *AIN2* is high, the motor will spin the other way.

All of this logic takes place in the **loop** method in the file *MainActivity.java*:

```
@Override
protected void loop() throws ConnectionLostException {
    // make a dead off zone in the middle
    if (Math.abs(left_) < 0.2)
        left_ = 0.0f;
    if (Math.abs(right_) < 0.2)
        right_ = 0.0f;

    // make sure duty cycle never > 100%
    if (Math.abs(left_) > 1.0)
        left_ = 1.0f;
    if (Math.abs(right_) > 1.0)
        right_ = 1.0f;
```

```
    pwma_.setDutyCycle(Math.abs(left_));
    ain1_.write(left_ >= 0);
    ain2_.write(left_ < 0);

    pwmb_.setDutyCycle(Math.abs(right_));
    bin1_.write(right_ >= 0);
    bin2_.write(right_ < 0);

    try {
        sleep(10);
    } catch (InterruptedException e) {
        e.printStackTrace();
    }
}
```

The `loop` method uses two values for the left and right motors, held in the member variables `left_` and `right_`. Each of these is a number between -1.0 and +1.0, where -1.0 is spinning one direction, +1.0 the other, and 0 in the middle is stopped.

So, first there is a bit of conditioning of these values so that there is a dead zone in the middle of the control, where if the unsigned value (Maths.abs) is less than 0.2, then it is forced to be 0 to keep the motor stopped.

Similarly, there are also checks to make sure the range is not exceeded.

We then set the 3 control pins for each motor to make sure it goes in the right direction and at the right speed.

Finally, the call to `sleep` allows a 10 millisecond gap between settings of the motor.

The user interface for all this is encapsulated in the `RoverControlView` class.

The virtual joystick control handles all the touch events in the following method:

```
@Override
public boolean onTouchEvent(MotionEvent event) {
    x_ = (int)event.getX();
    y_ = (int)event.getY();
        int x1 = x_ - xO_;
        int y1 = y_ - yO_;
        float xf = (float)x1 / diameter_; // +- 0..1
        float yf = -(float)y1 / diameter_;
        float left = (float) (xf * cos135 - yf * sin135);
        float right = (float) (xf * sin135 + yf * cos135);

    if (event.getAction() == MotionEvent.ACTION_DOWN) {
        context_.setMotors(left, right);
    }
```

```
        invalidate();
        return true;
    }
```

The math here converts the X and Y coordinates into left and right motor powers by rotating the coordinates of the event 45 degrees and then passing them to the public `setMotors` method in the `MainActivity` class, where they can be accessed by the `loop` method that we described earlier.

Conclusion

That concludes not just this project but also the book.

I hope you have enjoyed learning more about IOIO and trying out some of these projects. You will find other resources and errata at the books website [*http://www.ioiobook.com*].

The author is always interested to hear about improvements to the code,or extensions to the projects, and you will find information on how to contact the author on the website.

About the Author

Dr. Simon Monk has a degree in Cybernetics and Computer Science and a PhD in Software Engineering. Simon spent several years as an academic before he returned to industry, co-founding the mobile software company Momote Ltd. He has been an active electronics hobbyist since his early teens. Simon is author of a number of hobby electronics books including *30 Arduino Projects for the Evil Genius*, *15 Dangerously Mad Projects for the Evil Genius*, and *Arduino + Android Projects for the Evil Genius*.

Have it your way.

O'Reilly eBooks

- Lifetime access to the book when you buy through oreilly.com
- Provided in up to four DRM-free file formats, for use on the devices of your choice: PDF, .epub, Kindle-compatible .mobi, and Android .apk
- Fully searchable, with copy-and-paste and print functionality
- Alerts when files are updated with corrections and additions

oreilly.com/ebooks/

Safari Books Online

- Access the contents and quickly search over 7000 books on technology, business, and certification guides
- Learn from expert video tutorials, and explore thousands of hours of video on technology and design topics
- Download whole books or chapters in PDF format, at no extra cost, to print or read on the go
- Get early access to books as they're being written
- Interact directly with authors of upcoming books
- Save up to 35% on O'Reilly print books

See the complete Safari Library at safari.oreilly.com

O'REILLY®

Spreading the knowledge of innovators.

oreilly.com

©2011 O'Reilly Media, Inc. O'Reilly logo is a registered trademark of O'Reilly Media, Inc. 00000

Get even more for your money.

Join the O'Reilly Community, and register the O'Reilly books you own. It's free, and you'll get:

- $4.99 ebook upgrade offer
- 40% upgrade offer on O'Reilly print books
- Membership discounts on books and events
- Free lifetime updates to ebooks and videos
- Multiple ebook formats, DRM FREE
- Participation in the O'Reilly community
- Newsletters
- Account management
- 100% Satisfaction Guarantee

Signing up is easy:

1. **Go to: oreilly.com/go/register**
2. **Create an O'Reilly login.**
3. **Provide your address.**
4. **Register your books.**

Note: English-language books only

To order books online:
oreilly.com/store

For questions about products or an order:
orders@oreilly.com

To sign up to get topic-specific email announcements and/or news about upcoming books, conferences, special offers, and new technologies:
elists@oreilly.com

For technical questions about book content:
booktech@oreilly.com

To submit new book proposals to our editors:
proposals@oreilly.com

O'Reilly books are available in multiple DRM-free ebook formats. For more information:
oreilly.com/ebooks

O'REILLY®

Spreading the knowledge of innovators   oreilly.com